Dedications

This is dedicated to the many people that voiced a need for a book like this, the people that wanted to hear more about the many health benefits of the soybean *and* an easy way to consume them in their *purest* form. Without processing, without additives or preservatives. ❀

WITH A LITTLE HELP FROM THE SOYBEAN

Table of Contents

Facts and Information .... Page 1
    Why the Soybean? .... Page 1

Areas of Health That Soy may affect .... Page 3
    Heart Disease and Cholesterol .... Page 4
    Cancer .... Page 4
    Osteoporosis .... Page 5
    Kidneys .... Page 6
    Alzheimer's Disease .... Page 7
    Menopausal Symptoms .... Page 8

Soyfoods .... Page 9
    Soy Milk .... Page 9
    Tofu .... Page 10
    Soybeans .... Page 10
    Soy Flour .... Page 11
    Miso .... Page 11
    Tempeh .... Page 12
    Texturized Soy Protein .... Page 12
    Soy Protein Isolates .... Page 13

Isoflavone contents .... Page 13

Appetizers .... Page 14

Main Dishes .... Page 22

Desserts .... Page 40

Tofu .... Page 54

"With a Little Help From the Soybean," contains information about the amazing soybean. This was written to shed some light on some very interesting facts that have presented themselves during the past years that I have used soybeans as an alternative. I wanted to know how and why it worked, and most of all, why is this information not made more public? I will share some facts, and include some ways to add soybeans to your diet. Most beans are nutritious, but the soybeans contain a higher amount of protein and fat than other beans and with a low count of carbohydrate. Soybeans get about 38% of calories from protein. Other beans get about 30% of calories from protein. Protein in soy is of the highest quality. The Food and Drug Administration, (FDA) and World Health Organization, (WHO) give soy protein isolates a score of one, the highest score available by them. Soy protein ranks on an equivalent level as protein from meat and milk products. About 40% of calories in soy come from fat. This part is used by consumers in the food industry. Most of this fat is unsaturated. Linoleic acid, (54%) is polyunsaturated. Oleic acid, (23%) is monounsaturated, Palmitic acid, (16%) is saturated. They are one of the few plant sources of omega-3 fatty acids. (Important for the nutrition of infants)

Dry soybeans can be stored in an airtight container for long periods of time before cooking. They don't have to be refrigerated until they are cooked. Initial preparation is quite easy. Prepare a weeks supply at a time. Storing them any longer after they are cooked, tends to change the taste and effect. One pound of DRY soybeans will yield about one weeks supply, if you are consuming a cup a day. (The dry soybeans expand when cooked.) Put about one pound of dry soybeans in a colander or spaghetti strainer, Rinse them, remove any milling debris or anything that doesn't resemble a soybean. Put them in a large bowl, cover them with water and let them soak overnight, at room temperature. (Or at least 7 hours.) In the morning, rinse them in the strainer or colander, put them in a slow cooker or crock pot, cover them with water, and simmer them on medium setting at least 4 or 5 hours, stirring occasionally. (Do not add seasonings.) They must be covered with water during the whole cooking time. Test for tenderness. If you can squeeze a bean between your pinky and thumb, (Where your strength is at its least) then they are done. Make sure they are soft before you remove them from the pot. Undone soybeans can be disruptive to the digestive process. Let them cool in the pot, then rinse them with cool water, drain and store in a NON METAL container with a tight fitting lid in the refrigerator. They are now ready to use in any of the recipes in this book. The nutrition value in each recipe stays the same, simply add the nutritional values of the soybean or tofu, that you have included, and you will arrive at the **total** nutrition picture. When cooking or baking with the soybean, it is best to use glass baking dishes and pans. Please remember that the recipes in this book are **adaptations** and may require a bit of experimenting until you achieve the result desired. Be persistent, this CAN be done. Some measurements and cooking times may need adjusting to accommodate individual stoves and ovens. If you are watching your fat, sugar or salt intake, you can make the necessary substitutions, such as fat-free, salt-free, and sugar-free products, to accommodate your needs. But, remember you may have to alter other parts of the recipes to balance and accommodate the changes accordingly. If you're in a hurry and can't prepare a soy dish, add your total days measurement of beans, all at once, to a bowl of soup, chili, Chinese take-out, macaroni and pasta sauce, etc. It is a very social food, and will accommodate to most dishes that you add it to, if you can't cook that day.❀

Facts and Information

Why the Soybean?

As we learn more about the many health benefits, through research and testing, the need for expanded ways to use the soybean is becoming necessary.

The soybean has much nutrition. It contains "complete protein," which means that all eight amino acids, needed for our health, is in this bean. It is also the ONLY vegetable food that has complete protein. The soybean is rich in B vitamins, calcium, and a great source of oil and dietary fiber. Soybean oil is rich in the two polyunsaturated fatty acids that we all need, linoleic and linolenic fatty acids.

Growing concern for personal health has dictated a need for a change in our diets. This change is the addition of soy. The Japanese have used soy in their diet since day one, and they are the longest living people in the world. Even their dogs live longer. This might be due to the soy they have eaten all these years. Japanese people also have the lowest rate of some cancers, suggesting that there may be a protection factor involved.

Many doctors and scientists are hard at work presently, to find out why and how this works. The credibility of this inexpensive ingredient is proven almost every day. It is reported by the American Heart Association and the American Cancer Society, that there is, health benefits associated with the use of the soybean. They have claimed that, soy, along with dietary changes and exercise, is important for the improvement of our health. The media often reports new findings in the area of health improvement, often with the soybean as the source. We are a generation that looks to the natural for most of our needs. The soybean has provided a natural answer to many health issues.

This book explains some areas tested. It allows an understanding of how this ingredient can be incorporated in your everyday diet, with hardly any problem at all. All that is required is a bit of creativity, to turn some of your present recipes into much healthier ones by adding or substituting with the soybean. How much easier can life be?

This book contains recipes for everything from applesauce cake to baked ziti. Some family members may be hesitant, but, maybe it is because they probably have not tried a recipe modified with the soybean. My five year old granddaughter, Amie, didn't think she would like anything made with the soybean. After she tried an oatmeal raisin cookie made with soy, she changed her mind. She said that they weren't too bad at all! If a picky eater like Amie likes them, then they MUST taste good.

In a recent article found in the newspaper this fall, a nutritionist forecasted that eggs may soon be spiked with genistein, (An anti cancer chemical found in the soybean) to make eggs more nutritional and to help them lose their bad reputation.

The public interest in the soybean is becoming widespread. It is fast becoming a part of today's diet, due to the results of testing being done. The important thing is to be able to distinguish, the proper sources for the optimum benefit. If a product label includes soy, it does not necessarily mean that it contains enough soy protein to be effective.

If you remember a while back, the media broadcasted the benefits of oat bran and its ability to lower cholesterol. Everyone rushed out to buy foods that mentioned oat bran on the label, no matter how much oat bran was actually present. Some manufacturers sprinkled oat bran on top of their bread and, suddenly, it became oat bran bread. They did not mention that you may have to eat half a loaf a day to benefit! I experienced this, and bought everything I could find with oat bran on its label. My result? I gained weight and my cholesterol did *not* go down. They failed to mention, that along with oat bran, you have to restrict your intake of fats, increase your exercise, and really stick with it. I did find, that, if I took pure unprocessed oat bran and made it into a hot cereal, much like oatmeal, I could achieve a lower cholesterol count. Oat bran had to be eaten every day to obtain a lower cholesterol count. But, that, along with more exercise and fewer fats, more fresh vegetables, fruits, and grains, less red meat, and less processed foods, is why it was *finally* successful. This theory is much like what is to be expected with the soy protein foods. I'm sure the food industry will attempt to mirror the oat bran saga. That is why it is up to the individual to do their homework and distinguish what foods contain enough soy to achieve the optimum benefit. There are information resources such as the "Soyfoods Center" in Lafayette, California, that has a vast data base, "SoyaScan," containing everything you ever wanted to know about the soybean and the foods associated with it. For a FREE catalog of publications and leaflet of favorite tofu recipes, send a stamped, self-addressed business-size, (No. 10) envelope to:

                     Soyfoods Center
                     P.O. Box 234
                     Lafayette, CA 954549-0234 USA

Obtaining the services of a nutritionist is also recommended, to modify the rest of your diet to adapt to the soybean and its foods.
Not all of the nutritionists are very well versed in the area of soy. But, if you do find one that can understand, encourage them to seek more knowledge for you.
"With a little help from the soybean," allows you to use the cooked soybeans to create your own custom soyfoods, If you don't have the time to create your own, there are many soyfoods available at your health food store.
Scientists predict that, as more studies take place, a clearer picture will be available. The media is reporting new findings as they happen. The most current test of interest is the one being conducted at UCLA. Results from this study may have some very interesting effects on how the public views the alternative approach.
As for myself, I know it works for me. It has been successful in more than one area of my health already, I am just waiting for the scientists to prove how and why.
The pages ahead contain the latest results of studies on soybeans and its products from around the world, as well as many different ways to include them in your diet.
The amount of servings in most of these recipes vary, the original recipes were designed to serve the standard 4 to 6 people. The addition of soybeans or tofu may increase the amount of servings. After you have made the dish you can make note of the servings that resulted and record it as such. ❀

Areas of Health That Soy may affect

A few years ago a group of researchers gathered to discuss the health benefits of the soybean and its many foods. Since then, there have been other gatherings, updating their findings after research. Included in the pages ahead are some findings that have taken place since that first gathering.

Areas of health that soy has affected:

Heart Disease and Cholesterol
Cancer
Osteoporosis
Kidneys
Alzheimer's Disease
Menopause

The above mentioned conditions have been directly affected by the use of the soybean and it's powerful nutritional properties. There are a number of different studies regarding the impact of using the soybean as an alternative to some prescribed medication being offered, or using the soybean in conjunction to some of the therapy being used.
One of the problems that has caused some concern is, the fact, that not many people can tolerate the soybean on a daily basis for extended periods of time. In order to be successful, the soybean consumption *must* be consistent. There are many foods made with soy, available in today's marketplace which may make the adaptation easier, and it offers variety and convenience. Or use recipes found in this book.
"With a Little Help From The Soybean," is the title of this book. Those particular words were chosen purposely, to remind us that, adding the soybean to the diet is only one part. The other keys to achieve the health benefits include additional steps. Such as, daily exercise. (Pick something that you will stick with.) One of the best exercises, that fit this requirement, is brisk walking. You can start out slow and build it up, until you can comfortably walk for at least forty-five minutes a day. Take your dog, you both can benefit from the fresh air and exercise. Restrict fat, salt and sugar consumption, decrease red meat and eliminate whole eggs from your diet. (Use egg whites or egg substitute) Include more fresh fruits, vegetables whole grains and limit the intake of processed foods.
The Japanese have used the soybean in their diet for many, many years. They are the oldest living people in the world and they have the lowest incidence of some cancers. Even their dogs live longer. But, when they come to America and live our lifestyle, they tend to suffer the same health problems that the Americans do. This seems to indicate, that, this longevity of life and low cancer rate is not all genetic, but possibly the result of a combination of different lifestyle *and* diet. We should be taking notes and adapting to some of their ways, beginning with their *eating habits*. ✿

Heart Disease and Cholesterol

The issue of how soy protein can prevent heart disease and lower cholesterol has been studied closely by health professionals for many years. The latest reports, show that the isoflavones found in the soy protein are what helps to lower cholesterol. Soy protein without the isoflavones had no effect. It was suggested. that, soy enables cholesterol to be removed through the bile acid production. HDL, the "good" cholesterol, is increased when soy is consumed. If the HDL was quite low at first, it showed a greater increase when soy was added to the diet. The LDL, the "bad" cholesterol, was lowered, as well as the triglycerides. The figures are reported to be:

Cholesterol	Lowered by 9%
LDL	Lowered by 13%
HDL	Increased by 2%
Triglycerides	Lowered by 10%

Studies showed that, the higher the total cholesterol initially, the more significant the drop was. It seems that the properties in the soybean can direct the body to automatically balance the "good" *and* the "bad" cholesterol. The isoflavones in soy protein enhance a healthy environment for the heart, providing strength for the lining and vessels, reducing heart disease. This makes sense, because, the countries that eat soy regularly, seem to have fewer incidences of heart attacks.

Studies are still taking place in this area in an attempt to find out how and why this amazing change can take place. The message seems to be, soy *does* make a difference, but the diet and the lifestyle both must be altered to reap the benefits. Just including soy in your diet, without modifying your diet and increasing your exercise, may not allow for the improvement to take place. This is only one step of a multi step process.

Cancer

Cancer is not just one disease, but a combination of diseases in common. The normal cell is surrounded by cytoplasm. Healthy, normal skin and muscle cells work with each other for your bodies well being. The cancer cell is not as accommodating. It gets bigger and darker. The inside nucleolus becomes mixed up. It changes in shape and they begin to pile up on one another.

Cancer cells rob nourishment from healthy cells, growing more than normal cells, crowding them and pushing them out of the way, eventually starving them to death. This is when the abnormal cancer cells "take over" the whole area. The cancer will multiply and break through natural barriers and spread to nearby tissues. At first, this will be contained in one organ, but shortly, it will break through and spread to adjoining organs. (Extensions or local spread). Then they break through the walls of the blood vessels or other organs. (Distant spread or metastasis).

The isoflavone, genistein, found in soy protein, is believed to prevent the action of angiogenesis. Angiogenesis is the process that takes place when new blood vessels are created to feed the cancerous tumors. The isoflavones go to work, surrounding the cancerous tumor cells and containing them. This process keeps them from further destruction. (In some cancers)

Research in this area is ongoing, in the attempt to learn more about the unique power that the soy protein provides via genistein.

Genistein may prove to play a large part in the fight against the cancer that is hormone dependent. Tamoxifen, given to women who carry a high cancer risk due to genetics, and to treat primary breast cancer, is closely related to genistein.

Both Tamoxifen and genistein, act as anti-estrogens. Testing has proven that, both work on estrogen positive AND estrogen negative cells. They seem to know how to react to each need. Genistein is not toxic to cells and allows large amounts to be used in testing before it creates cell damage.

Other cancers involved in the soy-cancer testing are, prostrate, colon and endometrial cancer.

With soy in our diets, we can increase our protection against a disease that has ravaged *SO MANY* bodies already. This protective force is a combination of compounds that are found in the soybean. When science removes part of the soybean for testing or processing, it may be breaking up a system that, within itself, is able to achieve this protection as a whole. Scientists suggest we keep this system together to achieve the maximum benefit.

The National Cancer Institute was approached by a group researching soy and prostrate cancer, with the hopes of future funding. The National Cancer Institute turned them down. The reason for their reluctance, was, the inability to find people that would stay on a soy diet for at least a year. These researchers were already doing this, however, the National Cancer Institute was not persuaded.

Books like this one and the many different soyfoods available on the market may provide some assistance in this area, indicating that the soy diet can be appealing to the taste buds, without too much trouble, and allowing longer dietary studies.

Osteoporosis

Osteoporosis is thinning of the bones, increasing the risk of breakage and fractures. Many older women experience this condition. Some women of postmenopausal status, claim, "they don't need hormone replacement therapy, and have breezed through menopause without event." Osteoporosis may be taking place and these women are not even aware of it until a bone breaks.

Isoflavones found in the soybean, seem to mimic estrogen. It has been known to restore the estrogen levels in some postmenopausal women without the risks of the traditionally prescribed Hormone Replacement Therapy. A postmenopausal woman's estrogen levels must be at a level close to where it was before menopause in order for her body to absorb and metabolize calcium needed for the protection from osteoporosis. If not, the calcium she consumes will not be of much help because, it may not reach some of the necessary points. (Bones) Due to the elevated risks associated with the hormone replacement therapy offered today, many women choose not to replace their estrogen. This leaves them open to many health problems, one of which is osteoporosis.

The soybean is a natural food that can help. (It also contains calcium.) Tofu, a food derived from the soybean, made with a calcium based process, is rich in calcium. There are some foods that are high in calcium, such as spinach and beet greens, but, because they are high in oxalate, (An acid found in many plants) the calcium is not absorbed as readily as the calcium in milk. Some well absorbed foods are:

1 c. cooked, Broccoli	175 mg (calcium)
1 c. cooked, Navy beans	140 mg
1 c. baked, Vegetarian beans	125 mg
8 oz. Orange juice (calcium fortified)	300 mg
1/2 c. firm, Tofu	250 mg
10 dried figs	250 mg

It is suggested that women should consume at least 1,000 mg of calcium per day.

There is a similarity between the soybean isoflavone and the anti-osteoporosis drug, ipriflavone. This drug is used to promote better bones and fight osteoporosis.

One study claims that isoflavones, may help fight osteoporosis much like estrogen therapy does, but, without the risk.

Bone Densometry Tests were conducted at the beginning and at the end of this particular test period. The test charted the results of postmenopausal women eating foods that contained 55 mg of isoflavones, 90 mg of isoflavones, and a group that ate foods which didn't contain isoflavones. All groups consumed 950 mg of calcium a day. None of the groups were aware of the amount of isoflavones that their particular group was consuming. After six months, the groups that were consuming 90 mg of isoflavones daily, showed a small INCREASE in bone density AND bone mineral in the lumbar spine. The other groups showed no difference in bone density. (It is NEVER too late to start)

Kidneys

Kidneys are a very important part of our bodies. We often take them for granted. We don't realize the complex job that these organs do until we experience a problem.

Many of us consume twice as much protein as we need. Some of that protein comes from beef, chicken and fish. Beef makes the kidneys work the hardest, next is chicken and then fish, in that order. Over taxing our kidneys with large amounts of animal protein can lead to kidney disturbances and disease. When large amounts of this protein is consumed at one meal, the kidneys have to work extra hard and long, increasing their blood flow, to increase urine production, to remove the waste material.

Some recent tests suggest, that, if you suffer from diabetes and are at risk for kidney disease, over the age of fifty, and have high blood pressure, excessive animal protein consumption will almost always lead to kidney damage. When you consume large amounts of SOY protein in one meal, the kidneys do not have to over tax themselves and their job is a very smooth and easy one.

It is suggested that when trading animal protein for soy protein, it may offer us protection from disease. According to researchers, studies are being held presently, to understand the extent of improvement the soybean has to offer in this area of health.

Alzheimer's Disease

Colleagues at John's Hopkins revealed that there may a link between estrogen and Alzheimer's disease. The study was designed to identify the area of the brain affected by Alzheimer's disease and explore it. The study group involved 514 perimenopausal women that were tracked for 16 years while on estrogen replacement. The study revealed a 54% reduced risk.

A Baltimore study contains the strongest evidence of the protein effect of estrogen against the disease:

In vitro, estrogen acts as a trophic factor for cholinergic neurons. Cholinergic depletion is the most prominent neuro-transmitter deficit of Alzheimer's disease. Women and men have estrogen receptors throughout the brain, even in some areas that are in a state of disorder. Estrogen seems to be necessary to maintain the integrity of the section of the hippocampas, that is associated with memory changes, pertaining to aging and Alzheimer's disease. Phytoestrogen, is a plant form of estrogen found in the soybean. It has the capabilities to replace estrogen levels in menopausal and post menopausal women. This allows the protection of traditionally prescribed estrogen replacement preparations, but without the risks involved. Study results in this area are not complete at this time. However, it is believed that the isoflavones in soy can accomplish almost the same protection, naturally, as its manufactured counterparts. This seems to explain why the Japanese have the longest life expectancy in the world, and the soy protein that they have always consumed *is* useful in fending off aging.

Menopausal Symptoms

There are many studies taking place to identify and define effects that soy may have on menopausal and postmenopausal women. In recent testing, it was suggested that the isoflavones, which is a plant form of estrogen (Phytoestrogen) found in the soybean, has the ability to mimic estrogen. In premenopausal women, it is said to provide protection from breast cancer. If a woman has an ample supply of estrogen, it acts as an anti-estrogen. If the woman is postmenopausal, and has a low estrogen count, it acts as a pro-estrogen. This amazing compound always seems to know what to do and where to go. This is an important fact. The postmenopausal woman should replace her estrogen to remain healthy and ward off the ravages of menopause. However many are reluctant due to the health risks involved when taking the usually prescribed replacement therapy that is available. Some of the preparations that are available are labeled "natural." The only thing "natural" about one particular preparation (Premarin) is that, it is derived from PREGNANT MARE URINE. The side effects that occur to some women challenges the credibility of the purpose. The animal rights activists are taking measures to examine the inhumane treatment of these mares, and are distributing information regarding these practices to the public. I read this information in a leading women's health magazine. (Women's Health Connection) I checked into this and found some pretty upsetting conditions happening at some of the horse farms located in the Dakotas that are supplying the urine for these preparations. I looked into this issue, mostly, because I doubted that the women in our society would knowingly purchase a product under such conditions. I was wrong. This takes place every day, but, all who take it are not aware of this. *BUT,* this is slowly changing, with the help of "Women's Health Connections," other leading health magazines and *now in this book,* every effort will be made to inform would be users of this preparation, before they consume it. The above information adds to the growing list in favor of naturally derived products.

The soybean has been known to reduce the severity of menopausal symptoms. This was was presented at the American Heart Association's meeting November, 1996.

The suggestion of soy protein as an alternative is still being researched further. There have been some results that have been promising. This issue will be clearer within the next two years, when more test results will be available. I don't know how it works, but, I, and many other women, are using the soybean for this purpose with good results. A practicing surgeon, delivering a lecture on breast cancer, revealed that she would never resort to the hormone therapies being offered today. She is planning to use an alternative. She claims that she has seen too many women on her operating table that may have not been there if they chose a natural alternative for hormone replacement. This was enough proof for me, *AND* a major factor in my decision to use the soybean.

Dr. Wulf H. Utian, M.D., Ph.D,, director of Reproductive Biology at Case Western Reserve University in Cleveland, " cautions the power and drug-like properties of isoflavones." The doses should be monitored by a doctor when used for medical purposes.

Soyfoods
Soy Milk

Soymilk can be used in place of most recipes that call for cow's milk. The unfortified version of soymilk contains high protein, B-vitamins and iron. The soymilks that are fortified, contain a source of calcium, vitamin D and B-12. Soymilk is an alternative for those who are lactose intolerant, because it does not contain the milk sugar lactose. Infants that have had problems adjusting to the commercially prepared formulas, often fare well when introduced to soymilk infant formula, because it can be digested easy. Fresh soymilk has been made in Japan and China for many years, by grinding soaked, cooked soybeans and extracting the dissolved soymilk from the beans.

In our country, soymilk is sold in non-refrigerated cartons, or in refrigerated plastic containers, found in the health food stores and supermarkets.

Soymilk comes in unflavored and flavored form. Reduced fat content versions are also available. The non-refrigerated package can be stored on the shelf for many months, but, once it is opened, it will stay fresh for only five days. It is also available in powder form that is mixed with water. It can be stored in the refrigerator or freezer. Once it is reconstituted with water, it must be refrigerated and will last about five days.

This can be used instead of evaporated milk in many recipes. Pumpkin pies and custard pies will have a lower fat content when made with soymilk. Use soymilk with your cereals, add it to pancake, waffle and crepe mixtures. For a low cholesterol, low saturated fat effect, use it in your cream sauces and soups. Or, you can just use it as a plain drink or make a great shake using tofu, frozen yogurt and fruit.

Nutrition	Soymilk (8 oz.)	Low-fat (8 oz.)
Calories	140	100
Protein	10 gm	4 gm
Fat	4 gm	2 gm
Carbohydrates	14 gm	16 gm
Sodium	120 mg	100 mg
Iron	1.8 mg	0.6 mg
Riboflavin	0.1 mg	.11 mg
Calcium	80 mg	80 mg

Tofu

Tofu is made from the curd of the soybean. It is sometimes processed with calcium sulfate. It appears in the grocery stores, usually in the produce or dairy sections. It must be stored in the refrigerator, and it must be rinsed and covered with fresh water daily until you finish it. It will stay fresh for about five days. It can be frozen for up to five months. Tofu has the ability to take on the flavor of anything that it is mixed with.

4 oz.	Firm Tofu	Soft Tofu
Calories	120	86
Protein	13 gm	9 gm
Carbohydrates	3 gm	2 gm
Fat	6 gm	5 gm
Cholesterol	0	0
Saturated Fat	1 gm	1 gm
Fiber	1 gm	-
Sodium	9 mg	8 mg
Iron	8 mg	7 mg
Calcium	120 mg	130 mg
% calories from fat	45	52
% calories from protein	43	39
% calories from carbohydrates	10	9

Soybeans

Nutritional Values 1/2 cup (cooked)

Calories	149
Protein	14.3 gm
Total Fat	7.7 gm
Saturated Fat	1.1 gm
Unsaturated fat	6.6 gm
Carbohydrates	8.5 gm
Crude Fiber	1.8 gm
Calcium	88.0 gm
Iron	4.4 mg
Zinc	1.0 mg
Thiamine	0.1 mg
Riboflavin	0.3 mg
Niacin	0.3 mg
Vitamin B-6	0.2 mg
Folacin	46.2 mg

Soy Flour

Soy Flour is a fine powder that is made with roasted soybeans. It comes in comes in a natural or full fat variety or in a defatted form with the oils removed via processing. Defatted soy flour has more protein present. If you replace 1/4 of regular flour with soy flour you will give your baked goods more protein. These flours must be refrigerated.

3 1/2 oz	Full Fat	Defatted
Calories	441	329
Protein	34.8 gm	47 gm
Fat	21.9 gm	1.2 gm
Fiber	2.2 gm	4.3 gm
Calcium	188 mg	241 mg
Iron	5.8 mg	9.2 mg
Zinc	3.5 mg	2.4 mg
Thiamin (B-12)	.41 mg	.7 mg
Riboflavin	.94 mg	.25 mg
Niacin	3.29 mg	2.61 mg

Miso

Miso, which is derived from the soybean, is the ingredient that gives Asian cooking its character. It is a smooth paste that is is aged with salt, a grain, (Such as rice) a mold culture, and soaked in cedar vats. This process takes up to three years, to achieve the salty, flavoring ability. It is used in dressings, sauces, soups and marinades. It lasts many months, if refrigerated.

2 Tbsp.	
Calories	71
Protein	4 gm
Fat	2 gm
Carbohydrates	9 gm
Calcium	23 mg
Iron	1 mg
Zinc	1.25 mg

Tempeh

Tempeh is a cake of soybeans that is similar in flavor to mushrooms. It is made with soybeans, grains and a piece of already processed Tempeh. (Starter) It is wrapped in banana leaves and fermented for up to 24 hours. Tempeh lasts months in the freezer, and can be kept fresh in the refrigerator for 10 days. It tastes great when marinated in barbecue sauce or lemon, and then grilled. It is also used in pasta sauces, soups, and can be made into a sandwich spread.

4 oz.

Calories	204
Protein	17 gm
Carbohydrates	15 gm
Calcium	80 mg
Iron	2 mg
Zinc	1/5 mg

Texturized Soy Protein
(Also known as TSP or TVP)

TSP is an extender that is used instead of, or along with, meat. It is soy flour that has been processed to change the structure of the protein fibers. It is a dried product that comes in chunks or granules, which have to be reconstituted with boiling water. It can be kept in an air tight container at room temperature for quite a few months. Once reconstituted, it must be refrigerated and used in the next few days.

1 cup

Calories	120
Protein	22 gm
Fat	0.2 gm
Carbohydrate	14 gm
Calcium	170 mg
Iron	4 mg
Sodium	7 mg
Zinc	2.7 mg

Soy Protein Isolates

Soy Protein Isolates are the defatted flakes that result when the oils and husks are removed from the soybean. When the protein is removed from the defatted flakes, *those* results are soy protein isolates. They contain the largest amount of protein, (90 per cent) and are the most highly refined soy protein of all soyfoods. It has been a protein source that has been used in many foods as an extender, main ingredient or analog. It is used in health food supplements, formulas, and special athletic supplements. It contains amino acids, which are necessary for our maintenance and growth. It contains nine amino acids that our bodies need and don't produce. Eating plant foods will provide the one amino acid that is not as readily available.

Calories (1 oz.)	95
Protein	22.60 gm
Total Fat	0.95 gm
Saturated fat	0.15 gm
Polyunsaturated fat	0.61 gm
Monounsaturated fat	0.19 gm
Carbohydrate	2.10 gm
Crude Fiber	0.07 gm
Calcium	50.00 mg
Iron	4.00 mg
Zinc	1.10 mg
Thiamine	0.05 mg
Riboflavin	0.03 mg
Niacin	0.40 mg
Folacin	49.30 mg

Some Soyfoods and their isoflavone content

1/2 c. cooked soy beans	35 mg
1/2 c. soy flour	50 mg
1/2 c. tofu	35 mg
1/2 c. low-fat tofu	35 mg
1 c. soy milk (reg)	30 mg
1 c. soy milk (low-fat)	20 mg
1/2 c. Beef-Not (TSP)	122 mg
1/4 c. Nutlettes cereal (dry)	62 mg
2 Tbsp. roasted soy butter	60 mg
1/2 c. Tempeh	35 mg
1/2 c. Miso	35 mg
Take Care beverage powder	35 mg

APPETIZERS

Cocktail Meatballs
Cheese Puffs
Stuffed Clams
Stuffed Mushrooms
Cheese & Sausage Puffies
Shrimp Dip
Pea, Bean Soup
Spicey Cheese Rolls
Tortillas
Mushroom Soup
Stuffed Apples
Liver Pate`
Tuna Pate`
Cheese Mold
Bean Dip
Cream cheese-Dill Dip

APPETIZERS

COCKTAIL MEATBALLS

1 lb. ground beef
1 c. cooked soybeans
2 eggs (or 4 egg whites, or equal amt. of egg substitute)
1 Tbsp. brown sugar
1 c. mashed potatoes
1 c. dry bread crumbs
1 tsp. salt (optional)
1/4 tsp. pepper
1/2 tsp. ground ginger
1/2 tsp. ground cloves
1/2 tsp. allspice
3/4 to 1 c. milk
1/2 c. soy flour (to coat)
2 & 1/2 pt.'s soy milk
no-stick vegetable spray

Puree the soybeans in the blender, mix with ground beef, eggs, brown sugar, mashed potatoes, bread crumbs, salt, pepper, ginger, cloves, and allspice. Add reg. milk, slowly so that the mixture does not become too moist. Shape into small balls, the size of walnuts, roll in flour. Coat pan with non- stick spray. Brown meatballs on all sides, pour the fat from the pan, and discard. Add soy milk, simmer for 30 minutes, until slightly thick. Serve this with tooth picks, or over white or wild rice.

CHEESE PUFFS

1/2 c. butter or margarine
1 c. Cheddar cheese
1 c. tofu
1/4 tsp. paprika and 1/2 tsp. cayenne pepper
1 c. flour
1 tsp. garlic powder

Mix softened butter or margarine, cream cheese and tofu in a large bowl. In a medium bowl, mix the remaining ingredients well. Mix the contents of both bowls together, shape into medium sized balls. Put on a cookie sheet, chill 3-4 hours. Bake at 350 degrees about 15 minutes.

APPETIZERS

STUFFED CLAMS

1 c. cooked soybeans
6 & 1/2 oz. minced clams, drained (save the juice)
1/3 c. margarine, melted
mince 1 small onion
3/4 c. seasoned bread crumbs
2 Tbsp. Worcestershire sauce
Paprika
Dash of regular pepper

Mash the soybeans, and mix with clams and melted margarine, onion, crumbs and Worcestershire sauce and pepper. Add clam juice you saved. Put this mixture on clam shells by spoon, sprinkle with paprika. Bake 15 min. at 350 degrees. Put under broiler for 1 minute. Serve with lemon wedges.

STUFFED MUSHROOMS

2 lbs. large mushrooms
1 & 1/2 c. bread crumbs
1 c. soybeans
1/4 Tbsp. grated Parmesan cheese
1 tsp. each of oregano & basil
1/4 Tbsp. minced parsley
1/8 tsp. each of garlic & onion powder
1/2 c. beef broth pepper

Chop the stems off mushrooms. Mash the soybeans, mix with crumbs, beans, oregano, parsley, cheese, and other spices. Stir in broth, put into the mushroom caps, Broil 3 inches from heat for 3 minutes, until lightly browned.

APPETIZERS

CHEESE & SAUSAGE PUFFIES

1 lb. sweet or hot sausage
1 c. cooked soy beans
1 lb. sharp Cheddar cheese (shredded)
3 c. biscuit mix (prepared)
3/4 c. water

Take sausage out of casing. Puree soybeans in blender. Mix soybeans with the sausage. Cook in a pan about 10 minutes, until the mixture is crumbly. Drain the fat, and chill. Mix this with the cheese, biscuit mix, and water. Form into small balls. Bake 15 min at 400 degrees on a slightly greased pan.

SHRIMP DIP

1 c. sour cream
1/2 package of cream cheese (3 oz.)
1 & 1/2 oz. tofu
2 tsp. lemon juice
1 package of dry Italian dressing mix
1 can of shrimp, rinsed and drained

Mix the cream cheese (which has been softened) and tofu with the sour cream. Stir in the lemon juice, dressing mix and the shrimp. After thoroughly mixed, let it sit for one hour. Chill in refrigerator, serve with celery, carrots and other raw vegetables.

APPETIZERS

SPICY CHEESE ROLLS

1 package (3 oz.) cream cheese with pimento
1 small package of Tofu
1/2 lb. Cheddar cheese
2 Tbsp. light mayonnaise
1/2 c. chopped walnuts
2 Tbsp. chopped parsley
1/2 tsp. minced garlic
Dash of Cayenne pepper

Leave the cream cheese out until it becomes room temperature. Grate the cheddar cheese. Mix the cream cheese, tofu, mayonnaise, nuts, parsley, garlic, and cayenne pepper. Blend thoroughly. Split into 2 separate portions, and shape into rolls about 1 1/2 inches across. Wrap each roll in foil and put in refrigerator overnight. When serving, lightly coat the rolls with paprika and cut into slices. Serve with crackers.

PEA BEAN SOUP

1 smoked Pork butt (With bone)
2 lb. dried split peas
1 c. of cooked soybeans
chop 3 onions
grate 3 carrots
Chop 1 stalk of celery
1 small can of Tomato sauce
2 med. potatoes

In 6 qts of water, simmer butt for 1 1/2 hours. Remove butt from water. Cool. Separate the meat from the bone and skim the fat from the top of the broth. Wash the peas and cook them in the broth until they are tender (1 1/2 hours) Chop the onions and celery and grate the carrots. Then add them along with the soybeans and tomato sauce to the peas and the broth. Just before serving, put the potatoes in a blender, on high, with some water and add them to soup for the purpose of thickening.

APPETIZERS

TORTILLA CHIPS

1 bag tortilla chips (medium size)
3 oz. Monterey Jack cheese
1 oz. Cheddar cheese
1 oz. tofu

Press tofu until most of the moisture is eliminated, chop into tiny bits. Shred both cheeses. Place tortillas on plate, sprinkle with cheeses and tofu. Put in microwave oven for 2 1/2 minutes on medium setting, or until cheeses and tofu are melted. Serve.

MUSHROOM SOUP

1/4 c. margarine
4 & 1/2 c. sliced mushrooms (fresh)
3/4 c. chopped onion
1/4 c. fresh parsley (chopped)
1 crushed clove garlic
1 c. cooked soybeans
3 & 1/2 c. chicken broth
1/4 c. flour

Simmer, mushrooms, onion, garlic and parsley in margarine until tender. Add chicken broth to this and bring to boil. Reduce heat and simmer 15 minutes. Put soybeans in blender until they are pureed. Mix flour salt, pepper and soybean puree with wire whisk until smooth. Add this, a little at a time, to mushroom mixture, stirring, until thickened. Don't bring it to a boil. Serve.

APPETIZERS

STUFFED APPLES

3 oz. cream cheese and 1 1/2 oz. tofu
1/4 c. onion, chopped
1/8 c. garlic, minced
1/4 c. parsley, chopped fine
4 medium apples

Soften cream cheese and tofu. With a mixer, blend until smooth. Core the apples, remove the insides, leaving a 3/4" shell Mix the cream cheese and tofu with the onion, garlic and parsley. Fill the apples with this mixture. Put them in the refrigerator for at least 3 hours. Cut in wedges.

LIVER PATE`

1 lb. chicken livers
2 Tbsp. margarine
1/2 c. tofu
3 Tbsp. light mayonnaise
2 Tbsp. lemon juice
1 Tbsp. chopped onion
10 drops Worcestershire sauce
1/2 tsp. salt
1/2 tsp. dry mustard
1/4 tsp. pepper

Simmer chicken livers in margarine in a covered sauce pan until they are completely browned. Remove from pan and mash with masher or put in blender. Add the remaining ingredients and stir until thoroughly mixed. Put in a mold or a bowl and chill for three hours. Make a bed of lettuce on a plate, invert the mold or bowl, and remove carefully. Garnish with chives or sliced hard boiled eggs, and serve with crackers.

APPETIZERS

TUNA PATE`

4 oz. cream cheese and 4 oz. tofu
2 Tbsp. chili sauce
2 Tbsp. each of chopped parsley and chopped onion
1/2 tsp. Worcestershire sauce
2 cans (7 oz.) tuna, drained

Soften and mix cream cheese and tofu until smooth. Mix with chili sauce, parsley, onion, and Worcestershire sauce. Slowly introduce the tuna, mixing until all is blended. Put mixture into a mold or small bowl, and leave in the refrigerator for at least 3 hours. Unmold. Serve.

CHEESE MOLD

1 envelope unflavored gelatin
1 c. chili sauce
1/2 c. cottage cheese
1/2 c. tofu
1/2 c. light mayonnaise
1/2 c. whipped cream

Dissolve gelatin by stirring in 3/4 c. cold water, over low heat. Mix the chili sauce, cottage cheese, tofu, and the mayonnaise together. Add the gelatin. Blend well. Slowly, add the whipped cream. Put into a quart size bowl. Put into refrigerator for at least 3 hours. Carefully unmold on a bed of lettuce or cabbage. surround with a circle of different sized and flavored crackers, celery and carrot sticks, raw pieces of cauliflower, broccoli, red and green pepper strips.

APPETIZERS

BEAN DIP

1 can of pork & beans (1 lb.)
1 c. cooked soybeans
1/2 c. America cheese, shredded
1/4 c. minced garlic
1 tsp. chili powder
dash of cayenne pepper
2 tsp. vinegar
2 tsp. Worcestershire sauce
4 slices crispy, drained & crumbled bacon

Mix all of the ingredients except bacon and cheese. Heat in casserole dish at 350 degrees for 30 minutes. Sprinkle with cheese and bacon and heat for 10 more minutes or until the cheese melts.

CREAM CHEESE DILL DIP

1 package cream cheese (3 oz)
1/2 c. tofu
1 Tbsp. chopped, green olives & pimento
1 tsp. chopped onion
1/4 tsp. dried dillweed
dash of salt & pepper
2 Tbsp. light cream

Soften the cream cheese and tofu until it is smooth. In a bowl mix the cream cheese, tofu, olives, onion, dillweed, salt and pepper. Gently fold in the cream. Put in the refrigerator and chill for at least 3 hours. Serve in a bowl with, raw celery, carrots, peppers, cauliflower, broccoli and crackers.
The above dip can be used to stuff celery sticks. Chop them into bite-sized pieces and serve on a tray. You can also core an apple, remove all the apple except for a 1/2 inch outer shell. Put the dip inside this shell, chill for 3 hours, and cut into quarters or eighths. Serve.

MAIN DISHES

Squash Casserole
Spinach, Rice & Beans
Eggplant, Stuffed
Chili Macaroni
Chicken, Rice & Bean Casserole
Piggies In A Blanket
Pierogies
Pasta Sauce
Quick Lasagna
Ziti & Beans
Mock Meat Loaf
Mock Meat Loaf (Spinach)
Ground Soybeans
Mock Burgers
Gravies
Catsup
Kebabs
Soy Pasta Dishes
Sauces
Spaghetti (Creole)
Stuffed Peppers
Mushroom & Soy Noodles
Celery & Elbow Bake
Hash
Bean Burritos
Beans (Crock Pot)
Barbecue Chicken (Stove-Top)

MAIN DISHES

PEAS, BEANS & MUSHROOMS

3 c. hot cooked rice made with chicken broth
1 c. cooked mushrooms (sliced)
1 c. cooked green peas
1 small onion chopped
1 c. cooked soybeans
2 Tbsp. butter
2 Tbsp. diced pimento
salt and pepper to taste

Put all of these ingredients together in oiled casserole dish. Bake 20 minutes at 300 degrees. Serve.

SQUASH CASSEROLE

6 c. of sliced squash (any kind)
1 c. shredded carrots
1/2 c. sliced onion
1 can cream of chicken soup
1 c. sour cream
8 oz. flavored, seasoned bread crumbs
1 c. cooked soy beans
1/2 c. melted butter
Mozzarella cheese (optional)

Cook the squash, onion and carrots in water for 5 minutes. Drain. Mix soup and sour cream, then mix with vegetables. Mash soybeans, mix with bread crumbs, pour butter over it. Put bread crumbs, beans and squash mixture, in casserole dish, top with cheese. Bake at 350 degrees 1/2 hour. Serve.

MAIN DISHES

SPINACH, RICE AND BEANS

1 lb. fresh spinach
1 c. rice (any variety)
1 c. cooked soy beans
mince 1 small onion
1 can of stewed tomatoes
Salt and pepper to taste
1 Tbsp. Olive oil

Wash and drain spinach, pat dry. Heat olive oil and saute onion. Add tomatoes, water, (amount recommended to cook rice) beans, spinach and rice. Simmer covered for 20 minutes. Serve.

EGGPLANT STUFFED

3 medium eggplants
1 c. cooked soybeans
1/2 c. olive oil
1/2 c. seasoned crumbs
1 & 1/2 c. chopped onions
grated Parmesan cheese
mince 2 cloves of garlic
1 (1&1/2 lb.) can of tomatoes (drained)
1/2 lb. ground beef

Simmer eggplants 10 minutes. Chill. Cut eggplants in half long ways, scoop out pulp, chop, set aside. Mash soybeans. In large pan saute, garlic & onions. Add pulp, tomatoes, beef and soybeans. Cook over low heat, 15 min., stirring occasionally. Add crumbs and fill eggplants with mixture. Place them in a greased pan. Top with Parmesan cheese. Bake half hour at 350 degrees, in preheated oven.

MAIN DISHES

CHILI MACARONI

1 small box elbow macaroni
1 lb ground beef and 1 c. cooked soybeans, mashed
2 large can tomatoes and 16 oz. tomato sauce
1 can kidney beans and 1 c. cooked soybeans
chop 1 medium onion and 1/2 green pepper
1 Tbsp. chili powder

Cook macaroni. Set aside. Mix ground beef and MASHED soybeans, brown in skillet with onion and green pepper. Add tomatoes, simmer 15 min. Put all in a 6 quart pot. Add the tomato sauce, kidney and soy beans and seasonings. Simmer 3 hours. Mix chili with cooked macaroni.

CHICKEN, RICE AND BEAN CASSEROLE

1 stick margarine
1 c. rice
1 c. cooked soybeans
1 can each, cream of celery soup, cream of chicken soup, cream of mushroom soup
1 soup can water
2 lb. chicken parts

Melt the margarine, add the rice, beans, all of the soups and the water. Put all of this in a greased 13 x 9 1/2 x 2 1/2 inch pan. Place chicken parts in this mixture, bake 325 degrees for 2 1/2 hours.

MAIN DISHES

PIGGIES IN A BLANKET

1 med. cabbage
1/2 c. rice partially cooked
1 c. cooked soybeans (mash them)
dice one med. onion
1 lb. ground beef
1 can tomato soup
1 egg or 2 egg whites, (or equivalent egg substitute)
1 soup can of water
mince 1 clove of garlic
2 Tbsp. margarine

Remove core from cabbage, par boil leaves. Cool. Trim thick ridges, separate leaves. Saute onion and garlic in margarine. Mix with meat, soybeans, egg, rice and seasonings. Starting with thick end of leaf, spread some mixture on each leaf, roll up. Place rolls in pot, pour tomato soup and water over cabbage rolls, cover, simmer 2 hours. Can be baked at 325 degrees for 2 hours.

PASTA SAUCE

2 c. tomatoes (fresh-cooked or canned)
slice one med. onion
4 Tbsp. chopped parsley
chop 2 stalks of celery
2 tablespoons of soy, safflower or veg. oil
1/4 c. cooked, mashed soybeans
1 Tbsp. whole wheat flour

Boil tomatoes, onion, celery, and parsley 20 minutes, drain. Brown flour in oil in pan over medium heat. Add soybeans and tomato to mixture. Cook, stir and serve.

MAIN DISHES

PIEROGIES

FILLING
1 lb. cottage cheese and 1 c. soybeans (mash them together)
mince 1 med. onion and 2 cloves of garlic
1 egg yolk or equal amount of egg substitute
Mix the above ingredients and set aside.

DOUGH
2 c. flour
2 tsp. butter
3 beaten eggs plus 1 egg white, or equal amt. of egg substitute
3 Tbsp. lukewarm water

Sift flour and salt, add butter. Add beaten eggs and warm water, thoroughly mix. Knead dough 10 minutes. Shape in a ball. Cut in 4 pieces. Roll each piece into 12 inch pieces. Cut that into 10 squares. Put a teaspoon of filling in each square, fold in half, pinch edges together. Put a few at a time, in an (uncovered) large pot of salted water. Boil 8 to 10 min. Serve with sauce or gravy.

MAIN DISHES

QUICK LASAGNA

8 oz. broad Lasagna noodles
1/2 c. Ricotta cheese
1/2 c. tofu
2/3 c. shredded Mozzarella cheese
1/2 c. grated Parmesan or Romano cheese
2 & 1/2 c. pasta sauce

Preheat oven 375 degrees. Cook noodles per instructions on package, drain. In a large bowl, mix noodles with cheeses and tofu. In a 2 quart loaf pan, put some pasta sauce to cover the bottom. Put half the noodle mixture on next, then half the sauce on top of that. Repeat. Bake 25 to 30 minutes.

ZITI AND BEANS

1 lb of ziti
1/2 lb. ground beef
1/2 c. soybeans mashed
chop 1 med onion & 1 med green pepper
3 Tbsp. butter
1 (12 oz) can of tomatoes
1 can of kidney beans
1 c. cooked soy beans

Saute onion and green pepper in melted butter. Add ground beef and MASHED soy beans which you have mixed together thoroughly, and brown. Add beans, tomatoes and seasonings. Simmer 1/2 hour. Cook ziti according to box instructions. Mix meat, soybeans, tomatoes and seasonings.

MAIN DISHES

MOCK MEAT LOAF

1 c. soy macaroni
3 c. of cooked soybeans
4 tablespoons of grated onion
1 egg
1/4 c. of pasta sauce.

Cook the macaroni (Al a dente, not soft) and drain. Put soybeans in a blender until chopped well. Put all the ingredients together with beaten egg. Mix. Grease loaf pan with oil. Put entire mixture in pan. Bake 40 minutes at 350 degrees in preheated oven. Serve.

MOCK MEAT-SPINACH LOAF

1 c. of fresh spinach, chopped up
2 c. of cooked soybeans
1 c. of celery chopped up
1 c. of flavored bread crumbs
Season with thyme, sweet basil, onion and garlic as desired
1 tablespoon of soy, safflower or veg. oil

Mash the beans with a masher and combine with all ingredients except the oil. Use the oil to grease a loaf pan and to brush the top when done baking. Bake in preheated oven for 30 minutes at 375 degrees. Top with tomato sauce or fresh sliced tomatoes. Serve.

MAIN DISHES

GROUND SOYBEANS WITH VEGETABLE FLAVOR

3 Tbsp. soy, safflower or veg. oil
1 onion, chopped very fine
2 vegetable bouillon cubes or 1 Tbsp. bouillon powder
1 Tbsp. soy sauce
1 c. tomato juice or sauce.
3 c. ground up cooked soybeans

Brown the onion in the oil. Add the bouillon broth, tomato juice or sauce and soy sauce. Stir while cooking for a few minutes. Introduce the ground soybeans, mix well. Cook over medium heat until most of the fluid is evaporated. Serve.

To make cooked, ground soybeans. You will need:

1 c. dry soybeans
A pinch salt (optional)
Enough water to cover the beans

Put beans into a container with a pinch of salt and soak them overnight. The next morning, drain, rinse and put in blender on the chop setting. (If your blender doesn't have that option set it at medium.) Chop the beans until they look like little pieces of corn in sauce. In a regular pan add 2 or 3 c. of water and cook until soft.

MAIN DISHES

MOCK BURGERS

2 c. vegetable flavored ground soy beans
2 Tbsp. soy, safflower or veg. oil
2 eggs
1 & 1/2 c. Melba toast, bread crumbs, or wheat germ

Add vegetable flavored soy beans to beaten eggs. Mix well with bread crumbs, Melba toast or wheat germ, form into a patty, roll in wheat germ or bread crumbs, brown in oil. Put on rolls, serve with gravy or tomato sauce on top.

GRAVY - BROWN

2 Tbsp. soy, safflower or veg. oil
2 Tbsp. whole wheat flour
1 c. cold water
1 Tbsp. soy sauce
(optional) salt to taste

Mix soy sauce, oil and flour and cook until brown. Stirring constantly, add the cold water and stir until thick. Season. For a different flavor, add 1 tsp. powdered vegetable broth.

GRAVY - MUSHROOM

1 (8 oz.) can mushroom soup
1 Tbsp. soy, safflower or veg. oil
1 & 1/2 tsp. flour
1 Tbsp. minced parsley

Warm up the soup. Mix the oil and the flour. Put it together with the hot soup and make a sauce. Sprinkle the parsley on and serve right away. (Add more milk or water as needed)

MAIN DISHES

BOUILLON, CHICKEN, BEEF OR VEGETABLE GRAVY

2 Tbsp. soy, safflower or veg. oil
1 Tbsp. onion, chopped fine
2 Tbsp. whole wheat flour
1 bouillon cube or packet
1 c. water
Season to taste

Cook the onion in the oil for a few minutes. Slowly introduce the flour while stirring. In a c. of hot water, dissolve the bouillon cube or packet. Add to onion and flour. Stir until thick. Season to taste.

EGG GRAVY

4 Tbsp. soy, safflower or veg. oil
1 egg, beaten until foamy
1 c. flour that has been browned (in oil)
2 c. soy milk

Heat oil in pan, when fairly hot, add the beaten egg, stirring till the egg pieces are browned. Add the browned flour, stir until you achieve a smooth texture. Slowly add milk while you are constantly stirring. Season to taste. Bring to a boil and serve.

MAIN DISHES

CATSUP

(This can go with your mock meat loaf and mock burgers)

4 quarts quartered ripe tomatoes
2 red peppers (sweet) cut into strips
2 green peppers cut into strips
1 c. celery cut in pieces 1 inch long
1 c. onions chopped
1 bay leaf
1 Tbsp. celery seed
1 Tbsp. coriander seed
2 c. fresh lemon juice (minus the pits)
1/2 c. brown sugar
4 to 6 canning jars (pint size)

Blend the first 5 ingredients until they become liquid. Put into large pot. (Due to the amount of vegetables and the space they take, you may have to blend the vegetables 1/4 amount at a time, depending on the size of your blender container) Put the celery and coriander seeds along with the bay leaf in a small muslin bag and add to the sauce. Simmer until the sauce has been reduced to half the original amount. At this time, add sugar and lemon juice, and cook another few minutes until it has reached the appropriate thickness. Remove muslin bag, squeeze gently, then discard. Sterilize 4 to 6 pint canning jars, and while they are still hot, seal the sauce into each jar right away. This can be kept in the refrigerator for many weeks.

MAIN DISHES

KEBABS

1 large onion
1 red bell pepper
1 small eggplant
1/2 lb. of zucchini
1/2 c. olive oil
1 1/2 tsp. salt
1/4 tsp. pepper
1/2 c. cooked ground soybeans
1/2 lb. ground beef
1 tsp. ground cumin
1 tsp. ground cinnamon
1 Tbsp. & 1 tsp. dried mint leaves

Cut onion into wedges. Heat oven to 400 degrees. Cut bell pepper and eggplant into 1" chunks. Cut the zucchini in half, length-wise, cut the halves into 1/2" slices. Mix oil, salt, and pepper in a baking pan, put the vegetables in and coat. Roast the vegetables about 15 minutes, until tender, stirring once. Take out of the oven and set aside. Mix the soybeans, beef, cumin, cinnamon and mint. Shape into 15 balls. Heat the grill. Alternate the soy-meat balls and vegetables when putting onto the skewers. Cook directly over the hot grill, turning twice, until the soy-meatballs are done, about 10 minutes.

MAIN DISHES

(Dishes Made With Soy Pastas)

CHEESE AND SOY ELBOWS

2 c. soy elbows
1 c. white sauce (see recipe below)
1 c. cottage cheese
1/2 c. flavored bread crumbs

Cook the soy elbows as directed on package. In a casserole dish, put one layer of elbows, then one layer of cottage cheese, repeat until you have used all of the elbows and cottage cheese. Cover all of this with white sauce, (see recipe below), and sprinkle the flavored bread crumbs on top. Bake in the oven at 350 degrees for 1/2 hour. Serve.

WHITE SAUCE

2 Tbsp. whole wheat flour
2 Tbsp. soy, safflower or veg. oil
1 c. soy or regular milk
1/4 tsp. salt (optional)

Put flour and oil in pan. Stir well. Add milk and salt and stir until thick.

- For Mushroom flavored white sauce; add 1/4 c. chopped mushrooms.
- For parsley flavor, include parsley to taste
- Brown Sauce; include soy sauce to the white sauce recipe to get preferred shade of brown.
- Egg Sauce; Take one egg, raw beaten, scrambled or hard boiled and add to white sauce.

MAIN DISHES

MUSHROOMS AND SOY NOODLES

1 can cream of mushroom soup
(or 1 can mushrooms)
2 c. soy noodles
1/4 c. soy or regular milk
whole wheat or flavored bread crumbs

Cook the noodles as directed. Place the cooked noodles in a casserole dish. Season to taste. Cover with the mushroom soup or canned mushrooms. Cover that with milk. Sprinkle with bread crumbs. Bake at 350 degrees for 1/2 hour. Serve.

CELERY AND ELBOW BAKE

2 c. cooked celery
2 c. soy elbows
1 & 1/2 c. white sauce
2 tsp. powdered vegetable broth
whole wheat bread or toast crumbs

Cook elbows as directed. Stir celery and elbows together. Add white sauce. Put in a casserole dish. Sprinkle with bread crumbs. Bake 20 to 30 minutes at 350 degrees.

MAIN DISHES

SPAGHETTI, CREOLE STYLE

2 c. soy spaghetti
2 c. vegetable protein food (cutlet, steaks, etc.)
dice 1 med. green pepper
dice 1 med. onion
2 c. cooked tomatoes
4 Tbsp. olive oil

Cook spaghetti as directed. Chop up protein food. Simmer onion and green pepper in oil in a medium pot. Add vegetable protein, simmer until brown. Put tomatoes and spaghetti in the pot, season to taste, cover and simmer 15 to 20 minutes.

STUFFED PEPPERS

6 medium green peppers
1/2 lb ground beef, mixed with 1 c. cooked soybeans
1/3 c. chopped onion
2 cloves garlic, chopped
dash salt and pepper
1 can tomatoes (1 lb.)
3 c. water
1/2 c. uncooked rice
1 tsp. Worcestershire sauce

Cut the tops off of peppers remove seeds & membrane and wash. Mash the soybeans. In a medium bowl, mix the ground beef, soybeans, rice, salt, (optional) pepper, and Worcestershire sauce. Stuff the peppers with the mixture. In a large pot, put tomatoes and water. Place peppers in pot. Keep peppers covered with liquid, simmer until rice is tender. (3 hours.) Option: When fully cooked, stand stuffed peppers in broiling pan, sprinkle mozzarella cheese on top and broil until cheese is melted.

MAIN DISHES

HASH

1 c. coarsely ground cooked beef
1 c. cooked, chopped, soybeans
1 c. coarsely ground potatoes
1/4 c. chopped onion
1/4 c. chopped parsley
1 tsp. salt, dash of pepper
2 tsp. Worcestershire sauce
(1) 6-oz. can evaporated milk
1/4 c. bread crumbs
1 Tbsp. melted margarine

Mix the first 8 items gently. Place into 1 qt. casserole dish. Mix the bread crumbs and margarine and sprinkle on top. Bake at 350 degrees for 1/2 hour.

SOY BEAN BURRITOS

1 lb. ground beef
1 c. cooked soybeans
1 pkg. taco seasoning
1 med. pkg. cheddar cheese
1 can refried beans
flour tortillas

Mash soybeans, mix with ground beef, and taco seasonings, brown in skillet. Add refried beans. Put mixture into tortillas, sprinkle with cheese. Bake at 350 degrees, 10 - 15 minutes, in a greased baking dish, or until the cheese melts.

MAIN DISHES

BEANS IN THE CROCK POT

1 can kidney beans (drained)
1 can lima beans (drained)
1 can garbanzo beans
1 can butter beans
1 c. cooked soybeans
1 can pork & beans
1/2 lb. ground beef
1 large onion

SAUCE: (simmer 15 minutes)

1/2 c. catsup
1/2 c. barbecue sauce
1 tsp. dry mustard
1 tsp. chili powder
dash Worcestershire sauce

Simmer ground beef with onion, strain grease from it. Mix all the beans together. Simmer bacon, drain the grease except 2 Tbsp. Add ground beef and onion to the beans, mix well. Add the bacon with 2 Tbsp. grease, stir.
Place in a greased casserole dish and bake 1 hour at 350 degrees, or put in a slow cooker and cook all day long. Serve this with rice, noodles or fresh bread.

You can stretch your usual meat dishes with mashed soybeans. (Burgers, meat loaf, stews, etc) Doing this will increase the protein factor as much as 15%.
Just add 1/3 c. of soybeans to 2/3 c. of meat to arrive at one full cup.

MAIN DISHES

BARBECUED CHICKEN ON THE STOVE

1 chicken
1 bottle barbecue sauce
1 10 oz. bottle of ginger ale
1 c. soybeans

Cut chicken into desired pieces. Put in a pan that has been sprayed with non-stick coating, on top of the of the stove. Mash the soybeans. Mix soybeans with barbecue sauce, put in blender and mix well. Pour sauce and ginger ale over the chicken. Simmer on stove for 1 hour or until desired tenderness is reached. You now have barbecued chicken without using the oven or the grill. You can compliment this with assorted mixed vegetables and twice baked potatoes, or rice.

DESSERTS

Carrot-soy Cake
Date-nut Bread
Nutty-fruit Bread
Pie Crusts
Lemon Pie
Pumpkin Pie
Chocolate Pie (Easy)
Soy Cookies
Wafers
Oatmeal Cookies
Fig Cookies
Peanut Butter-soy Cookies
Wheat Germ Cookies
Carob Brownies
Molasses-honey Cookies
Molasses-nut Brownies
Oat & Whole Wheat Bars
Sugar Cookies (Easy)
Soy Cake (Scratch)
Applesoyce Cake
Frosting
Date Pudding
Pumpkin Pudding
Cereal pudding
Butterscotch Pudding
Rice Pudding
Topping
Fudge

DESSERTS

CARROT-SOY CAKE

Sift together in a large mixing bowl:
1 1/2 c. flour
1 c. sugar
1 tsp. baking soda
1 tsp. baking powder
1 tsp. ground cinnamon (and or 1 tsp. nutmeg and ginger)
1 tsp. vanilla
2 c. chopped walnuts
2/3 c. oil (vegetable)
2 eggs, or 4 egg whites
1 c. shredded carrot
1 c. cooked soybeans

Puree soybeans in blender. Combine with rest of ingredients. With mixer, beat 2 minutes on medium speed. Divide the batter and pour into 2 greased floured bread pans. Bake at 350 degrees for 45 min. or, until center tests done. Cool for 10 min. and remove from pan. Cool thoroughly, cut and serve.

CREAM CHEESE FROSTING

1 & 1/2 oz. fat free cream cheese, and 1 & 1/2 oz. lite tofu (only 1 % fat)
1 Tbsp. butter, (softened)
1 tsp. vanilla
2 c. confectioners sugar, (sifted)

In a small bowl, mix on low, tofu, cream cheese, butter, and vanilla. Slowly add sugar, beating until fluffy. Apply frosting to a thoroughly cooled cake. (When in a pinch, combine tofu and cream cheese and add to prepared vanilla frosting)

DESSERTS

DATE NUT BREAD

2 Tbsp. margarine
1 & 3/4 c. whole wheat flour
1 c. chopped dates
1/4 c. soy flour
1 c. brown sugar
1 c. chopped nuts,(any kind)
1 c. soy milk

Blend the sugar and oil. Add nuts and dates, sifted flours, dissolved yeasts and soy milk. Beat together. Pour into oiled loaf pan. Let rise 1/2 hour in warm place. Bake at 325 degrees for 1 hour.

NUTTY FRUIT BREAD

1 & 3/4 c. white flour
1 egg, well beaten
1/4 c. soy flour
1/2 c. dates, chopped
1 packet dry yeast
1/2 c. raisins
2 Tbsp. soy or veg. oil
1/2 c. nuts, chopped (any kind)
1/2 c. brown sugar
1 tsp. orange peel, grated
3/4 c. soy milk

Sift the flours apart from one another. Measure and sift with yeast. Mix oil and sugar, egg, milk, fruit, and nuts. Add peel and dry ingredients, mix well. Bake 350 degrees 1 hour.

DESSERTS

PIE CRUSTS

1 c. white flour or whole wheat flour
2 Tbsp. soy flour
1/2 tsp. salt
6 Tbsp. shortening
3 to 4 Tbsp. cold water
Use more to coagulate dough if needed

Mix the flours and the salt. Cut the shortening in with a shortening knife or fork. Add the water until you can make a smooth ball. Put on floured board. Oil a sheet of wax paper, place on top of ball. Roll until thin. Line the pie tin. Bake at 350 degrees until brown. Enough for single 9" pie crust.

PIE CRUST (SOY)

3/4 c. soy flour
1/4 c. white flour or whole wheat flour
1/2 tsp. salt
5 Tbsp. shortening or margarine
3 to 4 Tbsp. cold water

Sift the flours and salt together. Cut in shortening or margarine with pastry knife or fork. Add water until you can make a smooth ball. Put on floured board. Oil a sheet of wax paper, put on top of the ball and roll until thin. Line 9" pie plate and bake at 350 degrees until it is brown.

DESSERTS

LEMON PIE

3 lemons
1 can sweetened condensed milk
4 egg yolks
1/4 c. soy milk, chilled
1 9" pre-baked pie shell
whipped topping

Make juice from the lemons and mix with egg yolks and milks until smooth. Pour into already baked pie crust, and chill in refrigerator for 2 & 1/2 hours. Add whipped topping and serve.

PUMPKIN PIE

1/2 c. sugar
4 Tbsp. browned flour
1 & 1/2 c. cooked pumpkin
1 Tbsp. molasses
1/8 tsp. salt
1/2 tsp. vanilla
1 & 1/2 c. hot, double strength soy milk

Stir sugar with browned flour. Add this to the pumpkin. Mix all of the ingredients and put in pie shell. Bake at 350 degrees for 15 minutes, turn oven to 400 degrees, and bake for 15 more minutes. If you put a tooth pick in the center and it comes out clean, then you know that it is done.

DESSERTS

SOY COOKIES

3 Tbsp. white flour
1 Tbsp. soy flour
1 & 3/4 Tbsp. sugar
4 Tbsp. soy, safflower or veg. oil or margarine

Mix ingredients. (add a little soy or reg. milk if too stiff) With a pastry tube, put on to a well oiled baking sheet. Bake at 375 degrees, for at least 10 minutes. Cool, remove from sheet and serve.

WAFERS

2 eggs
1 c. brown sugar
1 tsp. vanilla
3 c. ground cooked soybeans

Beat eggs until thick. Add sugar and vanilla. Puree soy beans in blender and add to the mixture. Use spoon to drop to oiled baking sheet. Use fork dipped in water to pat into 1/4 inch thickness. Bake at 250 degrees until brown. Remove from baking sheet as soon as you remove them from the oven. Serve.

DESSERTS

OATMEAL COOKIES

1 c. margarine or shortening
2 eggs
1 tsp. vanilla
1 c. brown sugar
3 c. oatmeal
1 c. raisins, chopped
1/4 c. soy flour
1 & 3/4 c. whole wheat pastry flour
4 Tbsp. soy milk or regular milk

Mix the margarine, oil or shortening with the brown sugar and vanilla. Beat the eggs, add them and the raisins to the mix. Sift flours, add them, and oatmeal to mix. Add milk and combine well. Drop mixture on greased cookie sheet by teaspoon. Flatten with knife or fork. Bake at 350 degrees until they are brown, keeping a watchful eye on them.

FIG COOKIES

3 eggs
1 c. oil or margarine
1 & 1/2 c. brown sugar
1 tsp. vanilla
1 c. chopped walnuts
2 c. figs, ground up
1/2 c. warm water
2 & 1/4 c. whole wheat pastry flour and 1/4 c. soy flour

Add margarine to sugar and beaten eggs. Mix in figs and walnuts. Stir well. Add water. Add sifted flours, combine well. Oil a cookie sheet, drop mix by teaspoonful. Bake 350 degrees until brown.

DESSERTS

PEANUT BUTTER SOY COOKIES

1/2 c. margarine
1 tsp. vanilla
1 c. brown sugar
2 eggs beaten well
1/2 c. peanut butter
1/8 c. water
2 & 1/4 c. whole wheat pastry flour
1/4 c. soy flour

Combine the margarine, sugar and eggs. Add the peanut butter, vanilla, water and all of the dry ingredients. Mix all together. Drop onto greased cookie sheet by spoon, flatten with a fork, horizontally and diagonally. Bake at 350 degrees until they are brown. (Keep a watchful eye.)

WHEAT GERM COOKIES

1/2 c. margarine
1 egg
1 c. brown sugar
4 Tbsp. soy milk or reg. milk
1/4 soy flour
3/4 c. whole wheat pastry flour
1/2 c. wheat germ
1/2 tsp. salt

Mix the margarine and sugar well. Add beaten egg, milk, vanilla, and then fold in the salt and sifted flours. Add the wheat germ. Use teaspoon to transfer to a well greased cookie sheet. If batter appears too thin, add more wheat germ. Bake at 350 degrees 30 minutes, until brown.

DESSERTS

CAROB SOY BROWNIES

1/2 c. margarine
1 c. brown sugar
1 separated egg
1 tsp. vanilla
1/4 c. white flour
1/4 c. soy flour
2 Tbsp. carob powder
1/2 c. chopped nuts (any kind)

Cream the margarine, sugar, add vanilla and beaten egg yolk. Add salt, flours, carob powder, and nuts. Fold well beaten egg white gently into mix. Oil and flour a square pan. Bake at 350 degrees for 20 minutes. Remove from the pan, cut them into squares, and let them cool. Serve.

MOLASSES HONEY COOKIES

1 c. margarine
1/2 c. honey
1 tsp. vanilla
grate the peel from 2 med. oranges
1 c. ground up oatmeal
1/2 c. molasses
1 c. coconut macaroon (fine)
1/4 c. soy flour
2 & 3/4 c. of whole wheat pastry flour

Mix all ingredients, except for the flours, very well. Sift all three flours together. Add sifted flours, one c. at a time, mixing thoroughly as you add. Drop mixture by spoon onto a greased sheet. Bake at 350 degrees 30 minutes or until brown.

DESSERTS

MOLASSES NUT BROWNIES

1/2 c. margarine
1/2 c. molasses
1/2 c. brown sugar
2 Tbsp. carob powder
1 egg
3/4 c. whole wheat pastry flour
1/4 c. soy flour
1/4 tsp. salt
1 c. chopped nuts (any kind)

Mix margarine and sugar until creamy. Add molasses and well beaten egg. Sift in all dry ingredients, add nuts. Pour mix in square, shallow pan. Bake 325 degrees for 40 to 45 minutes.

OAT AND WHOLE WHEAT BARS

2/3 c. margarine
1 c. soy milk
1/2 c. grated coconut
1/2 c. toasted sesame seeds
2/3 c. chopped nuts (any kind)
3 Tbsp. honey
2 c. pitted and halved dates
2 c. oatmeal, ground up
2 c. whole wheat flour

Mix all ingredients with soy milk to form a ball of dough to roll. Split dough in half, roll to 1/8 inch thickness. Put into 2 pans. Prick with a fork. Bake at 350 degrees until brown, 30 minutes. Watch edges, which may brown quicker than the middle. Cut into bars. Remove and serve.

DESSERTS

SCRATCH SOY CAKE

1/4 c. soy flour
1 & 1/2 c. white flour
2 generous Tbsp. cornstarch
separate 4 large eggs
1 & 3/4 c. sugar
1/2 c. water
1/2 c. oil (soy, safflower, or veg.)
1/4 tsp. salt
1 tsp. vanilla

Sift the flours before you measure. Measure, then sift with the cornstarch for six times. Beat the egg yolks with the water until thick and foamy. Keep beating as you add the sugar. Add oil salt and vanilla. Whip this all together until the oil has been thoroughly incorporated into the mixture. Do not over whip. Add all of the flours at one time. Gently fold in the beaten egg white. Pour mixture into a greased and floured, square, cake pan, or bundt pan. Bake at 300 degrees for one half hour, increase temperature to 350 degrees, and bake for 3/4 to 1 hour longer.

DESSERTS

APPLESOYCE CAKE

1 c. margarine
2 c. brown sugar
1 tsp. cinnamon
1 tsp. vanilla
2 separated eggs
1 c. thick unsweetened applesauce
1/2 c. mashed, cooked soybeans
1/4 c. soy flour
3 & 3/4 c. whole wheat pastry flour

Mix margarine, cinnamon and sugar until creamy. Add beaten egg yolks, soybeans, and applesauce. Sift all the dry ingredients together and add to the mixture. Beat the egg whites until they are stiff and fold them gently into the mixture. Bake in a greased floured cake pan at 350 degrees for 40 to 50 minutes. Test with toothpick in the middle. If it comes clean, cake is done.

DATE SOY PUDDING

2 eggs
1 c. soy or regular milk
2 Tbsp. honey
2 Tbsp. soy, safflower or veg. oil
1 c. chopped dates
1 c. soy Melba toast crumbs
1 tsp. lemon juice
a pinch of salt

Beat the eggs. Add the milk, honey, oil, dates, crumbs and lemon juice. Mix well. Put this all in a well oiled baking dish and bake for 30 minutes at 350 degrees.

DESSERTS

PUMPKIN PUDDING

6 Tbsp. honey
1/2 c. whole wheat flour
1/4 tsp. salt
1 c. cooked pumpkin
1 tsp. vanilla
1 & 1/2 c. soy or regular milk

Mix all ingredients together and bake at 350 degrees for 45 minutes. Top with orange juice.

CEREAL PUDDING

1 c. cooked soy cereal
1/2 c. brown sugar
2 slightly beaten eggs
2 c. soy or regular milk
1/2 c. chopped dates
1/2 c. chopped raisins
1 c. bread crumbs (unflavored)

Mix all of the ingredients well. Pour into well oiled baking dish. Top with bread crumbs and brown sugar. Bake at 350 degrees for 20 minutes, until almost solid. Serve with whipped cream.

DESSERTS

BUTTERSCOTCH PUDDING

4 Tbsp. brown sugar
4 Tbsp. cornstarch
1/4 tsp. salt
1 & 1/2 c. soy milk
3/4 c. honey
2 egg yolks
2 Tbsp. margarine

Mix sugar, cornstarch, and salt in pan. Slowly add milk and syrup, mix well. Cook, stirring constantly until thick. Beat the egg yolk slightly, and add this to mixture, a little at a time. Cook for 5 minutes. Remove from heat, add margarine, mix, and pour into individual dishes.

RICE PUDDING

1/2 c. uncooked rice
5 c. soy or reg. milk
1 tsp. salt
1/3 c. brown sugar
1 c. raisins
2 tsp. soy, safflower or veg. oil
1 tsp. vanilla

Wash and drain rice. Add milk and salt. Pour into casserole dish. Cook over low heat stirring until rice floats. Add sugar, oil, raisins, and vanilla. Mix, bake at 300 degrees until top is brown.

DESSERTS

TOPPING

1 c. water
1 c. soy milk powder
1/2 to 3/4 c. soy, safflower or veg. oil
1/2 to 3/4 tsp. vanilla
6 drops lemon juice

Mix water with soy milk powder in blender. Mix in vanilla. Add oil slowly until thick. Add lemon juice and chill. (1 tsp. lecithin will make a smoother mixture, using less oil).

FUDGE

1/4 c. butter
1/2 c. honey
1/2 c. carob powder
1/2 c. nuts
3/4 to 1 c. of soy milk powder
1 tsp. vanilla
chopped nuts (to roll them in)

Mix the honey and butter. Add the other ingredients and shape into bars about 6" long. Roll in the nuts. Chill in the refrigerator. Cut as you serve.

For a quick and easy snack, spread the cooked soybeans evenly on a cookie sheet that has been sprayed with a vegetable spray. Sprinkle with a dried vegetable seasoning or garlic. Bake in oven at 300 degrees for an hour, or until slightly brown, mixing the beans up every 15 minutes.

TOFU

Tomato Soup Casserole
Tofu, Fried or Baked
Sheperd Pie
Lasagna
Welsh Rarebit
Rice Casserole
Croquettes
Patties
Mushroom Sauce
Tofu Tomato Sauce
Sunny Potatoes
Twice-baked Potatoes
Scalloped Potatoes
Mashed Sweet Potatoes
Pancakes
Fruit Cocktail Salad
Tofu Strips (Steamed)

TOFU

Tofu, is made by curding soaked pureed soybeans with calcium compounds. These compounds add to the calcium content of tofu. This is a versatile food that has been used in the Asian countries for a long time. It is very easy to work with and with a bit of creativity, it can be incorporated into many recipes. The pages ahead will offer some very interesting ways to include tofu in your diet.

TOMATO SOUP CASSEROLE

2 Tbsp. soy, safflower or veg. oil
1 c. celery (chopped)
1/2 tsp. minced onion
1/2 tsp. minced garlic
2 Tbsp. chopped parsley
1 c. mashed tofu
1/2 c. tomato soup
1/2 c. ground walnuts
2 Tbsp. bread crumbs
1/4 c. sliced green olives with pimento
1 Tbsp. dry instant potatoes
1/2 tsp. lemon juice
1/4 tsp. powdered thyme

Saute in oil, the onion, garlic, and parsley in a pan. Simmer until translucent but, not brown. Mix mashed tofu with tomato soup. Add nuts, bread crumbs and olives. Add the sauteed vegetable mix, dry potatoes, lemon juice and thyme. Mix well and put in well greased casserole dish. Bake at 350 degrees uncovered for 45 minutes or until brown on top. Sprinkle top with parsley and serve with pasta sauce.

TOFU

FRIED OR BAKED TOFU

1 cake of tofu
1 egg, beaten
1 tsp. celery salt
1 c. bread crumbs
2 Tbsp. soy, safflower, or veg. oil
1 c. pasta or mushroom sauce

Cut tofu cake in half, length-wise, then across into 1/2" pieces. Dip each piece into beaten egg that has been seasoned with celery salt. Then roll in bread crumbs. Fry in small amount of oil until brown, or place slices in a well-oiled pan, and bake until slightly brown. Serve with pasta or mushroom sauce.

SHEPHERD PIE

1 Tbsp. soy sauce
2 Tbsp. soy, safflower or veg. oil
1 onion large, diced
1 tsp. powdered veg. broth
1 & 1/2 c. water
2 Tbsp. whole wheat flour
2 c. tofu
1 c. mashed potatoes
1 sm. can peas
1 sm. can creamed corn
1 sm. can green beans
1 Tbsp. parsley, chopped fine

Simmer onion in oil until brown, add water, soy sauce, and seasoning. Add flour and stir until thick. Add tofu and parsley. Put in a well greased casserole dish and top with the peas, 1/3 of potatoes. then corn, 1/3 potatoes, then beans and the rest of potatoes. Bake at 350 degrees until browned.

TOFU

LASAGNA

1 lb. Italian sausage
1 clove minced garlic
1 Tbsp. whole basil
(1) 16 oz. can tomatoes
(2) 6 oz. cans tomato paste
10 oz. lasagna noodles
2 eggs
1 & 1/2 c. Ricotta or cream style cheese
1 & 1/2 c. tofu
1/2 c. grated Parmesan cheese
2 Tbsp. parsley flakes
1 lb. mozzarella cheese, sliced thin

Brown meat, drain fat. Add next five ingredients, simmer 30 minutes, stirring often. Cook noodles until tender, drain and rinse. Beat eggs, mix ricotta with tofu, add them to rest of ingredients except mozzarella. Put 1/2 noodles in 13 x 9 x 2 inch pan, put 1/2 ricotta and tofu mixture on that, add 1/2 mozzarella, and 1/2 meat sauce. Repeat. Bake at 350 degrees 1/2 hr.

WELSH RAREBIT

1 c. shredded sharp American cheese and 1 c. of tofu
3/4 c. milk
1 tsp. dry mustard
1 tsp. Worcestershire sauce and a dash of cayenne
1 egg, well beaten

In a sauce pan, heat cheese, tofu and milk on low, stirring continuously, until sauce is smooth. Add next 3 ingredients. Stir. Serve with broiled tomato slices on toast.

TOFU

RICE CASSEROLE

2 c. brown, wild, or long grain rice (cooked)
1 packet powdered vegetable broth
1 c. tofu
1 c. soy milk or reg. milk
1/2 c. seasoned bread crumbs

Mix the rice with the salt and powdered veg. broth. Put the a layer of rice in the bottom of an oiled casserole dish. Put a layer of tofu on top of that, add another layer of rice and repeat until all of the rice and tofu are gone. Add enough milk to almost cover the contents. Sprinkle with bread crumbs. Cover and bake at 350 degrees for one hour.

CROQUETTES

1 egg, uncooked
1 c. tofu
2 Tbsp. soy, safflower or veg. oil
2 hard-boiled eggs
1/2 tsp. soy sauce
Pinch of salt
dash of sage
1 tsp. onion, chopped fine
1 c. cracker crumbs
minced parsley
tomato slices

Beat uncooked egg very well, mix rest of ingredients. (except the parsley and tomatoes) and shape into croquettes. Roll them in cracker crumbs. Put them in a well oiled pan and bake at 350 degrees for 20 minutes. Sprinkle with parsley and tomatoes. Serve.

TOFU

PATTIES

2 Tbsp. minced onion
2 Tbsp. olive oil
2 Tbsp. chopped celery
1 c. cooked brown rice
1 c. tofu
1 c. cooked green split peas
1/2 c. bread crumbs
1 Tbsp. soy sauce

Simmer onion and celery until brown. Add remaining ingredients except for the bread crumbs. Shape them into patties, roll them in the bread crumbs. Put them on a well oiled baking sheet and turn them so both sides are oiled. Bake at 350 degrees for 25 minutes or until browned.

TOFU AND MUSHROOM SAUCE

4 Tbsp. soy, safflower or veg. oil
1/2 c. tofu
1 bunch of green onions, sliced thin
2 stalks of chopped fine celery
1 small can of sliced mushrooms
1 Tbsp. soy sauce
3 Tbsp. flour (Whole wheat)
1 c. cold water

Place tofu in flat, oiled, casserole dish. Simmer onions and celery in oil until soft. Add mushrooms, soy sauce, flour, and water. Pour this over the tofu and let it marinate for 2 hours. Put in oven and bake at 350 degrees for 15 minutes.

TOFU

TOFU TOMATO SAUCE

2 c. cubed tofu
2 Tbsp. soy, safflower or veg. oil
1/2 Tbsp. minced onions
2 Tbsp. each of, minced celery and bell pepper
1 can tomato soup or same amount of tomato sauce

Brown the vegetables in the oil and add slightly diluted tomato soup or tomato sauce, and mix. Place tofu in casserole dish and pour the sauce over it. Bake at 350 degrees for 1/2 hour.

SUNNY POTATOES

4 medium potatoes
2 c. cooked carrots
2 Tbsp. butter
1/2 c. soy milk
1/2 c. tofu
2 Tbsp. chopped onion

Peel and cook potatoes. Saute onions in butter until translucent. Mix all ingredients and mash.

TWICE BAKED POTATOES

Cut baked potatoes in half, scoop out the potatoes and mash, put the potato skins intact, aside, Add 1/3 c. of tofu, 1/2 c. of minced onion and season with powdered veg. broth and salt. Moisten with soy milk. Mix well, put into baked potato skins. Sprinkle with bread crumbs and mozzarella cheese and bake at 350 degrees until top is brown.

TOFU

SCALLOPED POTATOES

8 c. peeled, raw potatoes
(sliced thin)
1/4 c. chopped onion
1 c. tofu
(chopped into small cubes)
1 10 oz. can cream of mushroom soup
1 10 oz. can cream of celery soup
1 c. soy or reg. milk

Place 4 c. of the sliced potatoes in a well greased 12 x 7 1/2 x 2-inch baking pan. Mix the rest of the ingredients, along with 3/4 tsp. salt, a dash of pepper, and pour half the mixture over the potatoes. Repeat the layers. Cover and bake at 350 degrees for 1 hour. Remove cover and bake for another 45 minutes longer.

MASHED SWEET POTATOES

Mash 6 cooked sweet potatoes
add 1 c. mashed tofu
3/4 c. brown sugar
1/4 c. butter

Mix well and put in buttered 1& 1/2 quart casserole dish. Bake at 375 degrees for 1/2 hour, add 1/2 c. of miniature marshmallows, bake 5 minutes more. Serve.

Steamed tofu strips or pieces can be added to any soup in place of noodles. Cook tofu and add to hot soup just before serving. Mash tofu, add salt, powdered, vegetable broth, and other seasonings. Mix, add chives, olives, or pimento to make a sandwich spread.
Toast 2 slices of soy bread. Add margarine. Broil, briefly, upside down on plate of sesame seed.

TOFU

PANCAKES

2 separated eggs
1/2 tsp. salt
1 to 1&1/4 c. soy flour
1 c. soy milk

Beat the egg yolks and add milk. Sift the flour with the salt. Mix the egg yolks and milk with the flour and the salt. Fold in the beaten egg whites. Pour onto greased hot griddle and cook.

FRUIT COCKTAIL SALAD

1 package (3 oz.) lemon gelatin
1 package unflavored gelatin
1 c. cold water
1 can fruit cocktail (1 lb.)
1/2 c. syrup from fruit cocktail
1 pint creamed cottage cheese
1 small package tofu
1 package cherry gelatin (3 oz.)

Fix the gelatin as package advises. Pour into slightly greased 9 x 5 x 3 inch loaf pan. Chill until set. Soften unflavored gelatin in cold water. Boil the syrup. Take away from heat, and gradually introduce the unflavored gelatin to the boiled syrup and stir until dissolved. Mix the cottage cheese, tofu and fruit cocktail together. Then add the unflavored gelatin. Spread this over lemon gelatin and chill until firm. Fix the cherry gelatin according to package. Chill to the point that it mounds when dropped from a spoon. Pour this over the cheese layer, and chill till firm. Remove from mold. Slice and serve on lettuce with mayonnaise, whipped cream or sour cream.

Some recipes are adaptations of ones found in *"Better Homes and Gardens New Cookbook."* I thank Meredith Book Publishers for the privilege. ❁

Sources

Proceedings from The 1st Annual Soy Symposium in Peducah, Kentucky. Update made public - November, 1996.

Second International Symposium on the role of soy in preventing and treating chronic disease. Brussels, Belgium. Update made public-September, 1996.

Ann Blish - Medical records and journals.

Bill Shurtleff - Soyfoods Center, Lafayette, CA

Constance Hegerford, Women's Health Connection

United Soybean Board

Kentucky Soybean Board.

Indiana Soybean Development Council.

Dr. Wulf H. Utian, M.D., PhD., CWRU